STEM in the MILITARY

GPS in Warfare

By Tanner Billings

Cavendish Square

Published in 2023 by Cavendish Square Publishing, LLC
2544 Clinton Street, Buffalo, NY 14224

Website: cavendishsq.com

Disclaimer: Portions of this work were originally authored by Judy Silverstein Gray and Taylor Baldwin Kiland and published as *The Military GPS: Cutting-Edge Global Positioning System*. All new material this edition authored by Tanner Billings.

Cataloging-in-Publication Data

Names:Billings, Tanner.
Title: GPS in warfare / Tanner Billings.
Description: New York : Cavendish Square, 2023. | Series: STEM in the Military| Includes glossary and index.
Identifiers: ISBN 9781502665485 (pbk.) | ISBN 9781502665492 (library bound) | ISBN 9781502665508 (ebook)
Subjects: LCSH: Global Positioning System --Juvenile literature. | Navigation --Juvenile literature.
Classification: LCC VK559.3 B55 2023 | DDC 623.89/3--dc23

Editor: Kate Mikoley
Copyeditor: Abby Young
Designer: Andrea Davison-Bartolotta

The photographs in this book are used by permission and through the courtesy of: Cover Africa Studio/Shutterstock.com; series art (blue/orange background) Troundless/Shutterstock.com; series art (gray background) Nickolay Grigoriev/Shutterstock.com; p. 5 ThomasDeco/Shutterstock.com; p. 6 roundex/Shutterstock.com; p. 7 MCLA Collection/Alamy Stock Photo; p. 9 (inset) wikimedia commons/File:John Harrison Uhrmacher.jpg; p. 9 (main) courtesy of the Library of Congress; p. 11 DVIDS/U.S. Coast Guard photo by Petty Officer 1st Class Alan Haraf; p. 12 courtesy of NASA; p. 14 Sharaf Maksumov/Shutterstock.com; p. 16 Air Images/Shutterstock.com; p. 18 (satellite) Inkoly/Shutterstock.com; pp. 20–21 wikimedia commons File:Usno-amc.jpg; p. 22 courtesy of AF.mil; p. 24 rblfmr/Shutterstock.com; p. 25 Astrid Gast/Shutterstock.com; p. 26 DVIDS/Photo and caption by Army Sgt. M; p. 30 DVIDS/U.S. Marine Corps photo by Lance Cpl. Brian Bolin Jr.; p. 32 DVIDS/ U.S. Air Force photo/Heide Couch; p. 33 (bullseye) Kraphix/Shutterstock.com; p. 33 (ground) Sensvector/Shutterstock.com; p. 33 (parachutes) DVIDS/U.S. Army photo by Sgt. Amanda Tucker/ Released; p. 33 (plane) Jason Swart/Shutterstock.com; p. 34 courtesy of Army.mil; p. 38 Courtesy Graphics/AF.mil; p. 39 maroke/Shutterstock.com; p. 40 Tarikdiz/Shutterstock.com; p. 41 DVIDS/ Courtesy Photo/Joint Task Force Guantanamo Public Affairs; p. 42 insta_photos/Shutterstock.com.

CPSIA compliance information: Batch #CWCSQ23: For further information, contact Cavendish Square Publishing LLC at 1-877-980-4450.

Printed in the United States of America

CONTENTS

INTRODUCTION

Today, anyone who has a smartphone probably has some sort of GPS in their pocket. Navigation apps, such as Google Maps and Apple Maps, use GPS, which stands for Global Positioning System, to help users get from one place to another. How does the GPS know where the user is? It uses satellites that orbit Earth at altitudes of more than 10,000 miles (16,000 kilometers). When you use a navigation app on your phone, it figures out how far away some of the satellites are. Using this information, the GPS can figure out your exact location, within about 35 feet (10 meters).

This might sound like a pretty complicated system just to help people find their way to the store. However, GPS is actually used for much more than just helping civilians get from place to place.

The U.S. Department of Defense launched experimental GPS satellites in 1978, and the system was fully operational by the mid-1990s. The military has many uses for GPS technology, such as to help determine the positioning and velocities of battlefield missiles and artillery projectiles. GPS can even be used to measure continental drift and track the movement of the polar ice sheets.

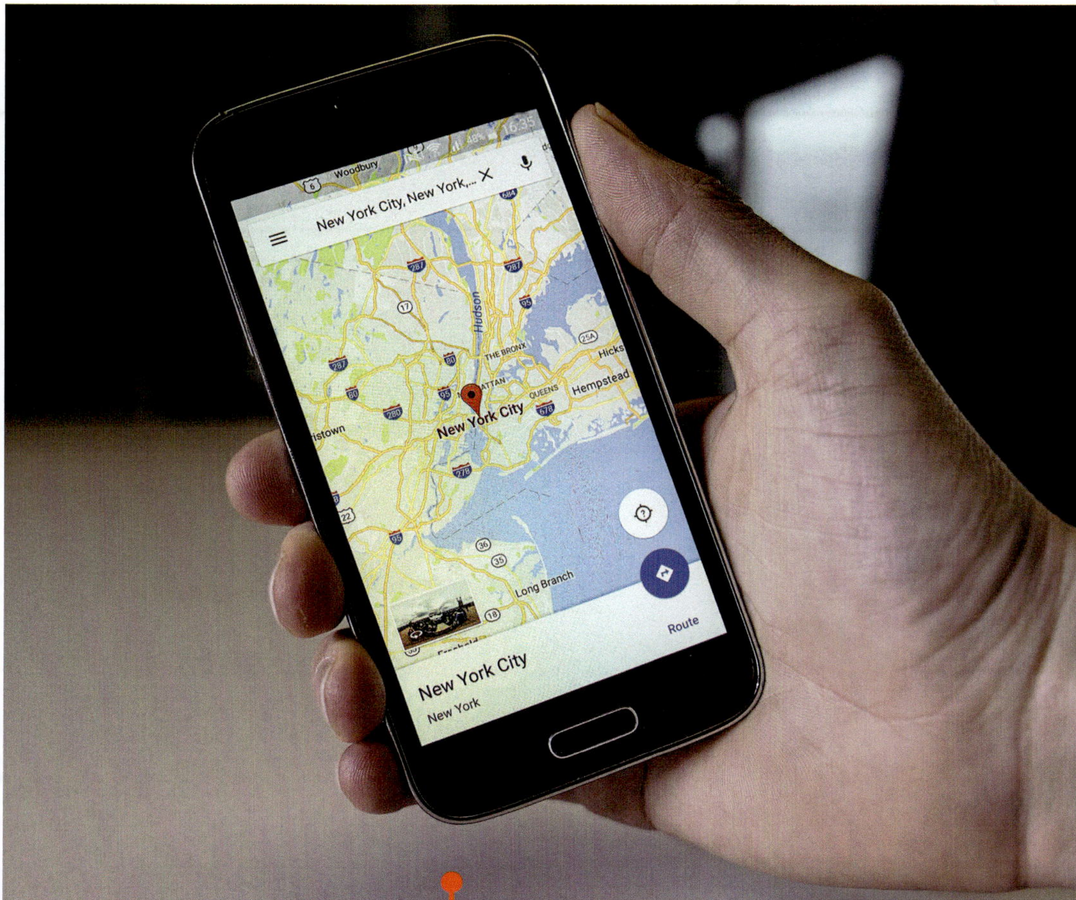

If you've used Google Maps, you've used GPS technology.

When you think of important tools used by the military, you may think of weapons, tanks, and aircraft. Indeed, the military needs all of these things in order to carry out its tasks. However, the technology that we can't always see down here on Earth, such as GPS, is also crucial for the military, especially during times of war. Read on to learn just how key of an element GPS is in warfare.

This well-preserved Viking ship is the Oseberg ship. While Vikings didn't have GPS, they did make some very important advances in navigational technology.

CHAPTER 1

NAVIGATION TECHNOLOGY HISTORY

Early civilizations traveling the seas relied on the sun, stars, wind, and waves to help them navigate. It's likely they also watched birds and sea creatures to learn about water currents and navigational hazards.

The Vikings traveled the seas between 750 and 1000 CE, using **primitive** tools. They were able to make their way to places such as Great Britain, Greenland, and North America without anywhere near the kind of technology we have today. Their calculations were based on when the sun rose and how high it rose during the day. Historians believe they passed down their navigation skills through storytelling.

Vikings used a tool called a plumb bob to determine the depth of water.

Although these primitive methods were impressive, their voyages were nowhere near as precise or orderly as voyages can be today with modern technology. The lack of preciseness also made for much more dangerous journeys.

Getting Better—Slowly

Calculating a location was vastly improved by developing a navigational system based on the stars and the solar system, called celestial navigation. It offered a way to determine position, but mistakes were common. A quadrant measured the angle of the sun and stars to estimate position. Miscalculations had interesting—and sometimes disastrous—outcomes. For example, the Pilgrims thought they were heading to what is now New York. However, the *Mayflower* landed the Pilgrims 220 miles (354 km) from their intended Hudson River destination.

The problem of developing ways to calculate **longitude** continued to prove perplexing. In the 1600s, Italian astronomer Giovanni Cassini timed the eclipses of Jupiter's moons, using a pendulum clock in Paris, France. This allowed him to successfully measure longitude and obtain a fairly accurate measurement of the longitude of Paris itself. However, he had solved the puzzle of determining longitude on land but not at sea.

In 1714, the British government supported the quest for mapping longitude at sea, passing legislation that offered a reward for anyone who could solve the puzzle. After experimenting for three decades, a London clockmaker named John Harrison developed an accurate and portable timepiece based upon how watches were made. Using a spring mechanism instead of a pendulum, he sought to accommodate the pitching and rolling of waves felt on a ship. During a nine-week Caribbean test voyage, the clock lost only 5 seconds, or 1.25 minutes of longitude. His clock reliably kept time at sea by mimicking the 24-hour daily rotation of Earth. Knowing the time that passed between one waypoint to another proved invaluable for making navigation more precise.

detailed design of chronometer

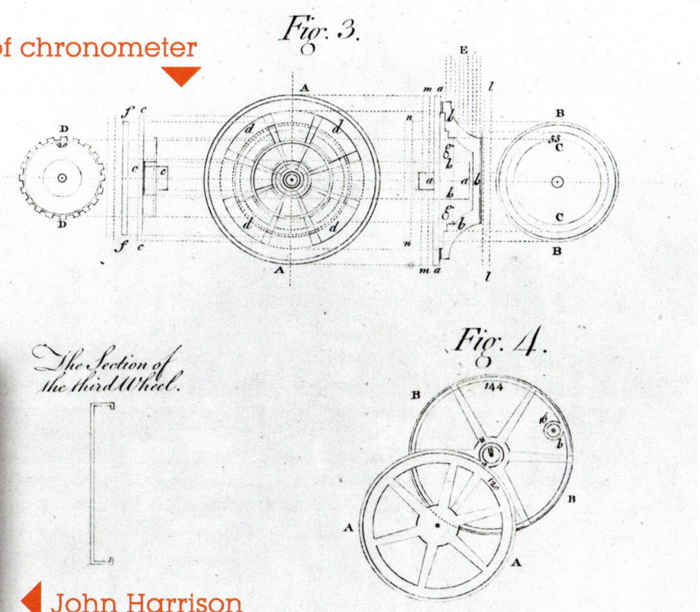

John Harrison

Harrison's invention was known as the marine chronometer.

As demand for shipping grew, so did the importance of finding a better system of navigation to improve safety at sea and on inland waterways. Ports and inland waterways maintained a system of marking and charting navigational hazards that continued to improve as overflights provided aerial information. Yet it was the launch of satellites, coupled with improvements in timekeeping, that revolutionized our ability to gather positional information more accurately, more quickly, and more reliably.

GPS Predecessors

Long range navigation (LORAN) was a land-based navigation system that determined lines of position by measuring the delay between when radio signals arrived from each of a pair of stations, located hundreds of miles apart.

The U.S. Coast Guard set up a chain of LORAN stations to provide navigational assistance to both ships and aircraft. By day, its range was at least 700 miles (1,126 km), and by night it reached a range of 1,400 miles (2,253 km). The lines on a screen were expressed as curves, and LORAN units converted this to degrees of **latitude** and longitude. Before it was phased out in favor of GPS, LORAN helped provide position information to ships and aircraft.

In 1958, the Advanced Research Projects Agency (ARPA) was created to ensure that the United States stayed at the forefront of technological advancements. ARPA played a role in the development of TRANSIT in the 1960s (which later became NAVSAT) and was a predecessor to GPS. Working closely with the Johns Hopkins Applied Physics Laboratory, TRANSIT became the first satellite-based position-locating system.

Now known as the Defense Advanced Research Projects Agency, DARPA often collaborates with engineers, physicists, mathematicians, and scientists from other government agencies.

LORAN was officially decommissioned in 2010. Shown here is a LORAN station from that year.

LOCUS
ASB-4
Antenna Splitter Box
POWER
LOCUS
LAD LORAN LRS IIID
12
STATIONS
LORAN-C MONITORING RECEIVER
STATUS
DELL
NEC
SEARCHLIGHT S/S 1600–1700Z
AccuSync LCD
REMOTE AUTOMATED INTEGRATED LORAN
RAIL CONTROL (Ver 3.2)

The Space Race

Amid the backdrop of tense relations between the **Soviet Union** and United States in 1957, *Sputnik 1* was launched by the Soviets. Weighing nearly 184 pounds (83.4 kilograms), the satellite orbited Earth once every hour and 36 minutes, transmitting radio signals strong enough to be picked up by amateur radio operators. Officially, the Soviet government claimed it was launched to correspond with a solar period considered ideal for launching satellites used for scientific study. However, the satellite was 10 times the size of the first planned U.S. satellite, not scheduled for launch until 1958. The U.S. military was caught off guard by the Soviet launch, thus setting in motion what the media, politicians, and historians referred to as "the space race." A year later, on January 31, 1958, the United States launched its first satellite.

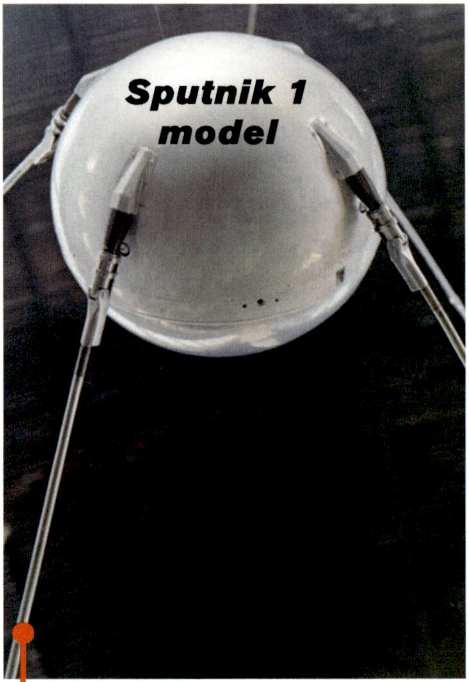

In Russian, *Sputnik* means "satellite."

Paving the Way for GPS

Following the October 1957 surprise launch of *Sputnik 1* by the Soviets, a group of American engineers, the military, and the space program began detailed observations of the Soviet space program.

Scientists tracking the satellite's radio transmissions soon realized they could pinpoint the location of the satellite using sound waves. When it was approaching, the sounds became more frequent. When it orbited away, the frequency of the sound waves decreased. This is called the Doppler effect. The scientists wondered: If we can determine the location of the satellite, can the satellite determine our location and other locations on Earth's surface? With that, the quest to develop a Global Positioning System (GPS) for the military began. In 1964, the first **Timation** satellite launched with **atomic clocks** on board, giving position-locating data a greater level of accuracy. Then, from 1978 to 1985, a group of 11 GPS satellites was launched, providing GPS-enabled data to the nation's military and intelligence communities.

GPS for the People

When the Soviets shot down Korean Airlines Flight 007, flying from New York City to Seoul, South Korea, on September 1, 1983, the reaction was swift. The Soviets claimed the commercial plane had veered into their air space without authorization. U.S. president Ronald Reagan realized the improved position-tracking features of GPS would provide greater safety for civilian aircraft so they would not enter unauthorized air space. He announced that when complete, the military's GPS system would be available for civilian use. His announcement was a game changer, and location-tracking information would become available for a growing number of military and civilian needs.

Operation Desert Storm

In 1993, the 24th **Navstar** satellite was launched, and the initial GPS constellation of satellites was completed. During Operation

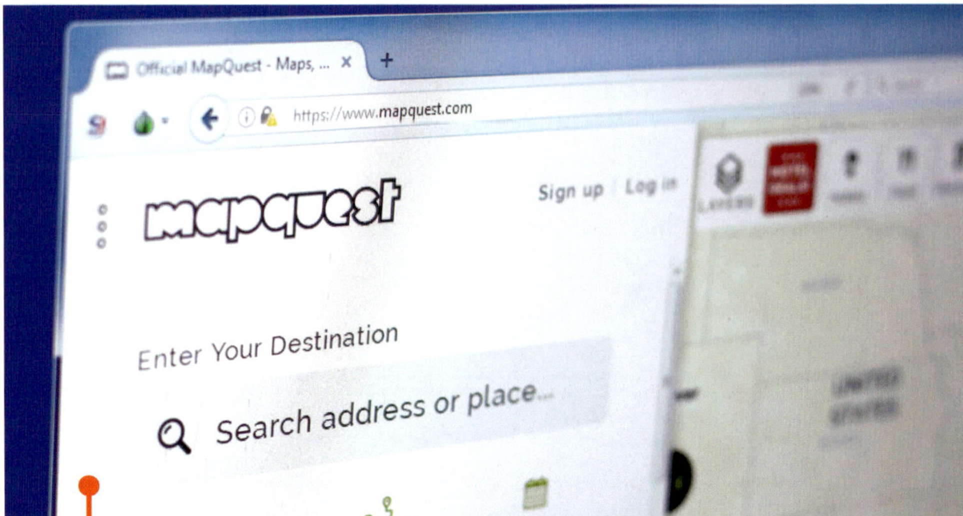

In 1996, MapQuest tapped into U.S. Census Bureau data to make maps accessible online. This was considered unimaginable when the technology was first introduced. Today, we have similar but more advanced tools such as Google Maps and Apple Maps.

Desert Storm (part of the Gulf War) in 1991, GPS made its combat debut. Soldiers were given handheld GPS receivers, enabling ground troops to determine their location within 100 feet (30 m). In the midst of the Iraqi and Kuwaiti deserts, and especially during sandstorms, when visibility was very poor for ground troops, this technology advancement offered a distinct military advantage.

Tank crews and drivers of all kinds of military vehicles—including food and supply trucks—relied on the accuracy of GPS to navigate the hostile terrain. Night maneuvers became easier—troops no longer needed to send advance scouts to guide the way. The technology also added a layer of agility to military operations. Troops could shift their plans because they were no longer dependent on stationary ground markers for location information.

GPS has also improved the performance of guided missiles, or smart bombs. Laser-guided weapons were first used on a large scale during Operation Desert Storm. The introduction of satellite guidance via GPS made the missiles even more accurate. Additionally, the Joint Direct Attack Munition (JDAM) and the Joint Stand-Off Weapon (JSOW), first introduced to the military in 1997, are both guided by an integrated inertial guidance system coupled with a GPS receiver. They can be successfully launched from fighter jet aircraft at a range of up to 15 nautical miles (27.78 km) from their target.

GPS in the Palm of Your Hand

As technology continued to advance, smaller and smaller GPS devices were made possible. This would be vital for the commercialization of GPS. The smaller the device, the more portable it is. Soon, people had their own GPS devices in their cars, and eventually it led to what we see today, where anyone with a smartphone has a GPS right in their pocket. Many people today can't imagine getting from place to place without this handy tool.

Join the Debate

1. When was the last time you used GPS to get somewhere? Could you have done it without GPS?
2. Why do you think GPS was so important during Operation Desert Storm?
3. How do you think the world would be different if GPS was not available for civilian use?

People use GPS for all sorts of purposes today, but many don't really understand how it works! These people are using GPS, along with a physical map, to help find their way on a hike.

CHAPTER 2

THE INS AND OUTS OF GPS

Even if you use GPS every day, you might not fully understand how it works. You type in your destination, click on the correct location, and your phone shows you how to get where you want to go. It may seem simple, but the technology that makes that all happen is a bit more complicated than just typing in the name of the place you want to go to.

How GPS Works

Within the GPS network, satellites transmit coded position signals and times as they orbit Earth. The signals are time-stamped as they are collected. Knowing the accurate time helps calculate the distance between signals received from each satellite and the GPS receiver on the ground. A GPS receiver is an electronic device that uses signals transmitted by satellites to determine its location (latitude, longitude, and elevation). The receiver provides position, speed, and time, allowing the user to determine their position.

The distance calculated is based on the time period between when the signal was sent and when it was received. Mostly, signals travel at the speed of light. Since satellites are positioned in space tens of thousands of miles above Earth, the calculations are a bit complex.

Trilateration

Distances from at least three satellites are received to determine the location of the original signal. This is called trilateration. Imagine you took a compass and traced circles around three separate locations. The point where the three circles overlap is the area where rescuers from the U.S. Coast Guard might look for boaters lost at sea.

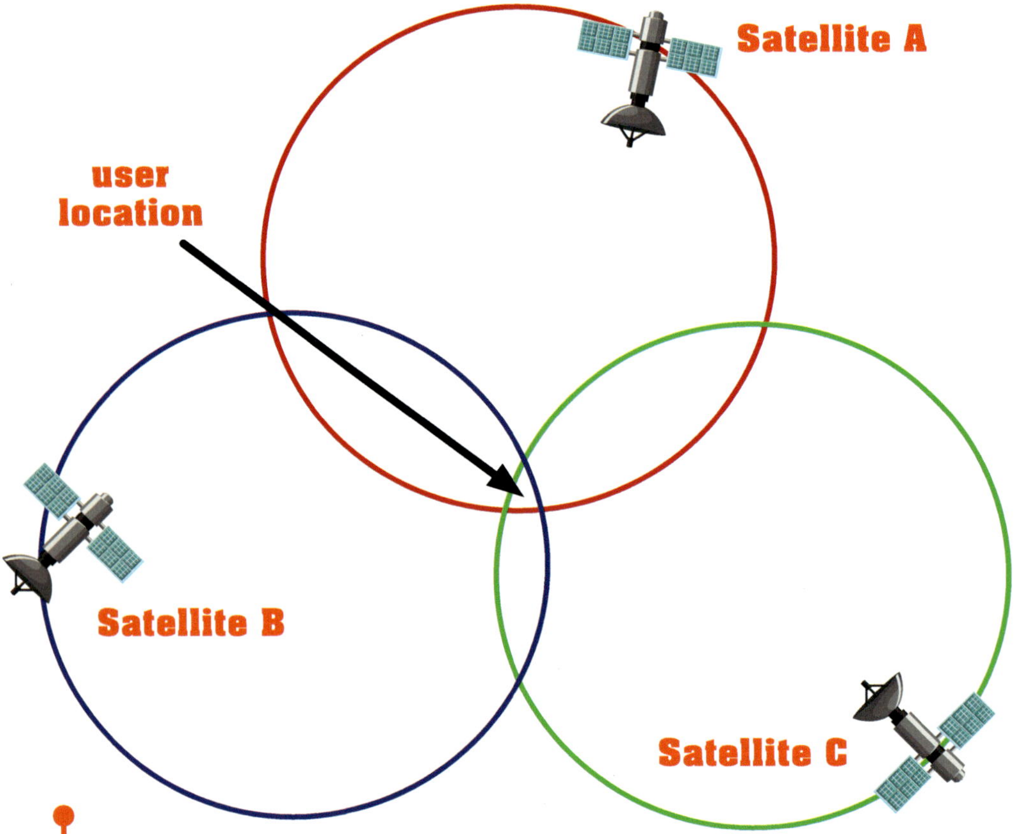

The point where all three satellites' circles overlap is where the GPS user is located. Often, four satellites are used for an even more accurate location.

It might also be where the U.S. Army would look for missing soldiers or U.S. Air Force crew members would look for paratroopers who didn't land where they intended. On a personal GPS device, information in the form of signals is received from several navigational satellites, and the device looks for the overlap. Then, your receiver, which has a clock synced with the GPS clock, determines your distance from each of the satellites by measuring the time taken for the signal to travel from the orbiting satellite to your receiver. It all happens in nanoseconds!

At least three satellite signals are received to determine the location of the original signal, but sometimes it's more. The energy sent from a satellite is traveling at the speed of light, so minor errors in calculation can result in larger location errors on the ground. Therefore, a minimum of four satellites is often used to reduce the error.

Today, navigational satellites can pinpoint a location within a few yards in any kind of weather, and their reliability is constantly being improved. By using a receiver and repeated signal measurements over several hours and then averaging the calculations, the position can be determined within a few inches. This same receiver technology is available in cell phones and equipment carried by U.S. ground and air forces. Now, we can determine our location anywhere in the world in fractions of a second!

Keeping the Satellites Working

The U.S. government owns the GPS. The program is paid for by U.S. taxpayers. GPS satellites are operated by U.S. military squadrons, or units, that are stationed at Schriever Space Force Base, which is located in Colorado Springs, Colorado.

Time Measurement

Ancient **mariners** knew that determining a precise location depended on accurately measuring the distance, or time, between two waypoints. Today, receivers measure time using a comparison between those imprinted on satellite signals, with the actual time recorded on a clock within the receiver. Time marks are controlled by highly accurate atomic clocks installed on board each satellite. Receivers also rely on signals from at least four satellites to maintain accuracy. For this system to work consistently, satellites need to transmit signals at precisely the same time.

An incredibly accurate master clock on Earth synchronizes all onboard satellite clocks at precisely the same time. Thinking back to Harrison's maritime clock developed in the 1700s, Greenwich Mean Time (GMT) remains the standard reference time for our planet. It is the time found in the town of Greenwich, located near London, England. In the United States, the official time for the Department of Defense is kept by the U.S. Naval Observatory in Washington, D.C.

The U.S. Naval Observatory (USNO) alternate master clock is at Schriever Space Force Base in Colorado. It serves as an alternate to the master clock in Washington, D.C.

However, the program also receives advice, guidance, and oversight from the National Executive Committee for Space-Based Positioning, Navigation and Timing, which is chaired by the president's deputy secretaries of defense and transportation.

GPS is incredibly useful, but the one drawback is that the satellites require constant maintenance and vigilance. They also need to be replaced as they age. The 45th Launch Support Squadron at Cape Canaveral Space Force Station in Florida is responsible for readying new satellites for launch

The GPS IIF-9 satellite is shown here being launched from Cape Canaveral in 2015.

11,000 miles (17,702 km) into orbit. It takes workers nearly three months to prepare a satellite for launch, and their work is conducted in two facilities at the base the space force uses to launch space missions. As of July 2022, the GPS constellation included a total of 31 operational satellites.

Now that so many people—both military and civilian—rely so heavily on GPS, it's more important than ever to keep the system working properly. The people responsible for maintaining the satellites have very important jobs. If GPS were to fail, there would be more trouble than not being able to figure out how to drive somewhere. Emergency systems would likely start to fail, with operators having difficulty locating many callers, and some military missions could be forced to come to a halt.

Join the Debate

1. Can you think of any drawbacks to GPS being able to determine a person's or object's location so closely?
2. Why is time measurement so important for GPS technology?
3. Do you think it's right that U.S. taxpayers pay for GPS? Why or why not?

Thanks to GPS, you can follow your food's trail straight to your door. In the military, GPS is used for more crucial purposes.

CHAPTER 3

GPS ALL OVER

Today, GPS is used for all kinds of reasons by all kinds of people. It's used by the military, governments, businesses, and individual people. It's used for commercial purposes, such as for tracking people's activity with fitness trackers and weather mapping. Farmers use GPS for crop planting, and wildlife experts use it to track animals' movements. Construction workers are able to closely follow survey maps while building roads thanks to GPS technology. When you use a delivery app such as Uber Eats or DoorDash to order dinner, GPS helps

Even farmers depend on GPS. These dairy cows are being tracked with GPS collars.

you follow your meal's exact location as it's being brought to your house. GPS makes location tracking possible in so many situations in daily life. In the military, the stakes are higher, so GPS accuracy is especially important.

GPS on the Front Lines

GPS satellites help soldiers with "situational awareness" on the battlefield, providing them with the location of friendly forces as well as hostile activity. In a war zone, this can mean the difference between life and death. Soldiers use a system called blue force tracking (named for a military term describing friendly forces). The GPS-enabled technology uses a linked network of vehicles, radios, satellite systems, software applications, and handheld devices. It allows individual soldiers to communicate with fellow ground troops. It's also helpful to track their positions in vehicles, aircraft, and command posts.

This soldier is programming a device that uses blue force tracking technology.

"Instead of having to run back to the vehicle or wait for grid coordinates, I can look down, see where my soldiers are and [...] make a decision of where to deploy them on the battlefield," says U.S. Army 1st Lieutenant Jason Kruger. He served with the 2nd Brigade, 1st Armored Division.

The U.S. Army often embeds a chip on each package or piece of equipment it ships so it can track and monitor its gear. Not only do radios and rifles include a form of GPS tracking, but some foreign militaries use embedded GPS-enabled chips to allow tracking of individual soldiers. The advantage is that real-time information provides a complete picture of what is happening, allowing split-second, life-saving decisions. In medicine, it can allow a patient to swallow a capsule containing a tiny camera with GPS and a light source, which maneuvers through the body, photographing a patient's internal organs. Radio signals provide information about the location of injuries or infection.

GPS at Sea

The U.S. Coast Guard taps into the remarkably precise data provided by GPS through its Rescue 21 system. The system has improved efficiency in gathering highly accurate position data and the success of search and rescue missions. It is also used to pinpoint buoy placement and to help map navigational hazards such as shoals, reefs, and sunken ships. The commercial fishing fleet uses GPS to improve safe navigation in ports and harbors, which are monitored by the U.S. Coast Guard.

GPS allows for more accurate underwater surveys, which can assist in marine exploration. Federal agencies such as the National Oceanic and Atmospheric Administration (NOAA), the

Department of Transportation, and the Department of the Interior use GPS for undersea mapping, bridge placement and building, climate study, and marine mammal tracking.

GPS on the Job

Since the U.S. Space Force, which is organized under the U.S. Air Force, operates and maintains GPS satellites, many career opportunities in the U.S. Air Force offer the chance to work with satellites and other space-based systems. GPS data allows pilots to calculate the amount of fuel on board and to plot a safe travel route. In military operations, GPS is used extensively for command, control, communications, computers, and intelligence. If you excel at GPS, you can work in intelligence, monitoring terrorist activity, geography and land analysis, remote mapping, security, and precision-guided weapons maintenance. Military maritime units rely on GPS to help in search-and-rescue operations and to increase the safety and flow of shipping traffic in ports. Military members may use GPS to forecast the weather, monitor foreign missile launches, and help recover aircraft or missing soldiers.

Not interested in the military but still want a job that uses GPS? Commercial pilots and other airline employees often work with GPS. Navigation apps such as Google Maps and Apple Maps need people to work on their products too.

Life Without GPS?

We rely on GPS technology for a lot, but what if it all went away? Today's military worries about cyberattacks as much as it worries about physical attacks. A cyberattack could take down the whole

GPS system, among other important technologies we rely on. While the military works hard to prevent this from happening, they also need to consider what they would do if it did happen. For this reason, navy sailors and coast guard officer candidates are required to learn about navigation using the stars, or celestial navigation. This might not seem particularly useful when we have technology that can navigate for us, but if the systems ever go down, we'll be glad to have people trained on celestial navigation.

Join the Debate

1. What do you think is the most important use for GPS?
2. Would you want a career that uses GPS? What type of job would you want?
3. Why do you think it's important to have navigational skills that don't involve technology?

While parachutes might not seem like the most advanced military technology, when combined with GPS technology, they make for a pretty important tool for servicing remote areas where troops are located.

CHAPTER 4

AIRDROPPING CARGO

Imagine ordering something online and having it dropped by parachute directly in front of your door. While that kind of delivery service isn't yet available to the average civilian, something pretty similar is available to troops in the military! The Joint Precision Airdrop System, or JPADS, is a dynamic tool in the world of resupplying troops in remote locations. The system uses parachutes and GPS to ensure supplies get to the exact location where they are needed.

Supplies typically dropped by JPADS include food, water, ammunition, and medical supplies used to resupply ground forces. The cargo is so important that the airlift occurs on schedule even if the weather conditions are bad. These drops are more reliable than delivery by helicopters and vehicle convoys and include fuel that the base depends upon to provide power and heat during long, snowy winters.

Setting the Scene

Picture the following scene from the time of U.S. combat missions in Afghanistan, which ended in 2021: A U.S. Air Force C-130 Hercules neared an American military base in a remote area, the huge aircraft all but invisible. Flying at an altitude of 17,000 feet (5,180 m), the plane was not only too high to be heard or seen from the ground, it was also well above the range of enemy weapons. This was extremely important to ensuring the success of its life-saving mission.

The C-130 Hercules has played an important role in getting supplies to military personnel.

Bundled supplies strapped to pallets underneath them were outfitted with a parachute and pushed out of the back of the plane as it passed over the designated drop zone at Combat Outpost Herrera, in east Afghanistan. Under normal circumstances, the parcel, driven by the subarctic climate and prevailing winds, could land a great distance from its intended target. However, a small, inconspicuous device attached to the pallet packed with vitally important supplies allowed the GPS to direct the bundle to the correct destination.

Once each supply bundle was launched, an **autonomous** guidance unit (AGU) began to direct the parachute. As it drifted downward, the AGU guided the bundle to the programmed point of impact. Its invaluable guidance system is similar to those used with a smart bomb and other direct-attack weapons.

"The bundles, when released from the aircraft at a previously calculated position, [guided] themselves to the drop zone" using a device attached to the parachute, explained U.S. Air Force Major Justin Brumley. The parachute was directed to its destination, easily negotiating mountainous terrain and other obstacles, "much the way a parachutist exacts a free-fall jump," Brumley said.

Combat Outpost Herrera sat 8,700 feet (2,650 m) above sea level. The remote military camp looked out over the surrounding mountains, trees, and local villages below. The accuracy that GPS provided the airdrops helped avoid the risks posed by jagged cliffs.

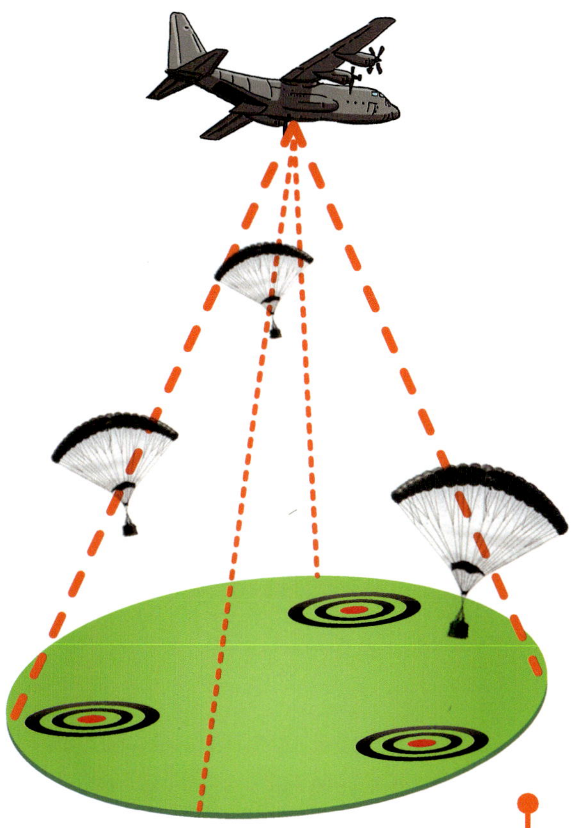

With JPADS, a single airplane can make multiple airdrops at once.

This photograph shows bundles of water being dropped to Combat Outpost Herrera in 2009.

Making It Possible

The combat airdrop missions made possible by the men and women of the 772nd Expeditionary Airlift Squadron at Kandahar Airfield, Afghanistan, delivered millions of pounds of vital supplies to troops at forward operating bases, like Herrera. From its inception in April 2009 to the end of that initial year, the mission dropped more than 2,000 computerized delivery system bundles, totaling about 4 million pounds (1.8 million kg) of supplies for the ground troops.

Army staff sergeant Denton Poe, a first sergeant at the outpost, was especially appreciative of the air-dropped fuel. "Without that unique tool," he said of the JPADS technology, "some of the stuff could come to a screeching halt here. Everything here runs on fuel—generators, heaters, vehicles. Without that fuel, a lot of stuff would come to a stop. Fuel is a lifeline here."

Humanitarian Airdrops

Not all the airdrops in Afghanistan were for the troops. Humanitarian airdrops provided food, water, clothing, and other basic supplies to local Afghan communities. "We know that those drops are helping people in the surrounding area," explained army captain James Speakes, deputy operations officer and officer-in-charge of tactics for the unit at the time. "Most of our drops are for the war fighter, but we also know that our mission here is to provide Afghans with assistance as well."

The airdrops reduced reliance on ground vehicle convoys, which was critical in remote outposts where roads may have been washed out or nonexistent. Because GPS provides precise position information and guidance, the air missions saved fuel, provided a higher level of efficiency, and improved the safety and security of ground forces.

The GPS-guided airdrops also had a great accuracy record. The "recovery rate" ("recovery" refers to a parachute parcel dropping near enough to its target to allow for its safe retrieval) improved to 97.2 percent after its first year of operation. This accuracy made believers out of even the most senior military personnel. "I was skeptical [of JPADS] at first," said army captain Brandon Kimbrel, a commander at Combat Outpost Herrera at the time. "There was one bundle way off in the distance," he recalls of one airdrop. "It was way off and we figured it was gone. Then, all of a sudden, it circled around and landed smack in the middle [of the designated target]. It was pretty cool to see."

Using GPS to Save Lives

After a day of fishing with friends in the Atlantic Ocean off the coast of Florida's Daytona Beach area, Billy Moody found himself in deep trouble. Grabbing the microphone of the vessel's marine radio, he shouted words every boat owner hopes to never utter: "Coast guard, Mayday. The boat is going down, we're sinking." The message was received at the U.S. Coast Guard's Ponce de Leon Inlet Station, but it left one huge unanswered question: Where exactly was the 22-foot (6.7 m) boat captained by Moody and occupied by five other teenagers?

"When we received the initial radio distress call," explained Sandy Ketchen, command duty officer at U.S. Coast Guard Sector Jacksonville, "the man didn't provide his full GPS coordinates." That **ambiguity** left a vast area of ocean to be searched by coast guard vessels, already about 90 minutes away from the area. Fortunately, Moody's mom had bought three emergency position indicating radio beacons (EPIRBs) for her teenage boater. The portable device emits a GPS emergency signal when it hits the water.

"Without the full position provided by the EPIRB, we would have been searching a 3,600-square-mile [9,320 sq km] area to find the vessel," Ketchen said. The coast guard arrived in time not only to save the crew but also to pump out and tow in the vessel.

GPS in the Sky

In the past, when you thought about GPS, you probably thought about how it was used on the ground, in cars and trucks. Maybe you thought about it being used in planes or helicopters too. Chances are, though, unless you already knew about JPADS, you didn't think about GPS being used in parachutes! This is just one of the many innovative ways the U.S. military utilizes GPS technology.

Join the Debate

1. How else, besides ways mentioned in this chapter, do you think JPADS could be used in everyday life?
2. Without GPS, would JPADS be able to carry out its job?
3. Without GPS, how do you think the coast guard could have found Billy Moody and his friends?

The NTS-3 is set to be the United States' next experimental navigation satellite. Expected to launch in 2023, it will likely pave the way for more advancements in satellite navigation.

CHAPTER 5

THE FUTURE OF GPS

It's amazing to think that because of some satellites in space, our phones can determine nearly our exact locations at pretty much any given moment. Despite this, there are still some times where GPS may not be exactly accurate. Certain structures, such as tall buildings, or weather events, such as extreme cloudiness, can interfere with or delay GPS signal transmissions. Being indoors, especially in basements, can make GPS work less accurately. In the military, accuracy is extremely important, so crews work tirelessly to ensure GPS is as accurate as possible. As more progress is made, there's no telling what capabilities GPS could have in the future.

Although it's probably rare, there may be times when your GPS thinks you are somewhere you aren't! This can happen when the signals are blocked.

Moving Forward

Experts work hard to make GPS better every day. This includes work on an

GPS OCX

The GPS Next Generation Operational Control System (GPS OCX) is a modernization effort being carried out by the U.S. government and Raytheon Intelligence & Space. Its aim is to make the ground control segment, or the ground-based stations charged with monitoring and updating GPS satellites, even better than it already is. The system, in development as of 2022, is aiming to make it possible to fly at least two times the number of satellites previously able to be flown. This will make for better coverage, especially in areas where signals are typically weak, such as those with many tall buildings or mountains.

This new system will be more cyber-secure than ever. Modernized equipment includes the latest methods of blocking GPS jamming, a process where a special device is used to interfere with signals, and GPS spoofing, which is when someone sends false GPS signals with a radio transmitter.

With GPS OCX, the U.S. government hopes to fly at least twice as many satellites as before.

early warning system for missile attacks on the United States and more GPS-enabled devices for deployed troops and the nation's allies. It also funds the development of GPS-enabled intelligence gathering.

Upgrades in technology mean greater accuracy for military units and greater precision for aircraft and vessels. It also means less money spent on fuel and locating those missing in action. According to a 2009 article in *Wired* magazine, the military has spent millions of dollars developing "tagging, tracking, and locating" (TTL) gear to assist in secretly tracking the position of people and items. The tiny chips can be embedded into radio-frequency identification (RFID) tags, sewn into fabric, or woven into paper to relay critical position-tracking information. In Afghanistan, some U.S. Army uniforms were equipped with Defense Advanced GPS Receivers (DAGRs), which could direct artillery toward targets.

You may be able to take classes in school to learn about GPS. Here, a student studies a GPS receiver with a military officer.

Today, the U.S. government has a multibillion-dollar program dedicated to modernizing GPS technology for both military and civilian use. These efforts include adding new civilian signals, enhancing cybersecurity, maintaining the performance of current satellites, and studying ways to create newer, more advanced ones. With the advent of NTS-3 and GPS OCX, the ways in which we use GPS will only continue to grow. Perhaps when you're an adult, you'll invent the next big thing in GPS technology—or maybe the thing that replaces GPS altogether!

If you work hard to learn about STEM subjects, maybe you'll help make some of the next big developments in GPS!

Join the Debate

1. How do you think GPS could be improved?
2. How do you think GPS will be different 10 years from now?
3. For all we know, in 50 years, GPS technology could be replaced by something entirely new. Describe what you think a new system that would replace GPS could look like. How will it work?

GLOSSARY

ambiguity: The quality or state of not having a clear meaning.

atomic clock: A clock that uses the measurement of the vibrations of atoms to give incredibly accurate standard measurements of time.

autonomous: Able to work on its own.

latitude: A distance north or south of the equator, measured in degrees. One degree of latitude covers roughly 69 miles (111 km).

longitude: A distance measured in degrees and marked by imaginary vertical lines east or west of the prime meridian.

mariner: A sailor.

Navstar: A network of satellites transmitting synchronized global positions.

primitive: Describes something from a very long time ago that is usually very unadvanced.

Soviet Union: A former communist country that stretched from eastern Europe to Asia and was also known as the USSR.

Timation: A satellite system that uses atomic clocks to help improve navigational accuracy.

FIND OUT MORE

Books

Hamilton, John. *Space Stations and Beyond*. Minneapolis, MN: ABDO Publishing, 2019.

LaPierre, Yvette. *Inside GPS*. Minneapolis, MN: ABDO Publishing, 2019.

Linde, Barbara M. *All About Road Maps and GPS*. New York, NY: Gareth Stevens Publishing, 2019.

Websites

GPS.gov
www.gps.gov
This official website from the U.S. government has information about the Global Positioning System (GPS).

Time and Navigation
timeandnavigation.si.edu
This website from the Smithsonian Institution has links to all sorts of information on navigation and its history. There is even a short video on how GPS works.

What Is GPS?
https://www.nasa.gov/directorates/heo/scan/communications/policy/what_is_gps
This article from NASA has lots of information on GPS and how it works.

Organizations

Federal Aviation Administration
800 Independence Avenue SW
Washington, DC 20591
www.faa.gov/
The Federal Aviation Administration is part of the U.S. Department of Transportation. It oversees GPS use in civil aviation. The organization's website has lots of information about satellite navigation.

U.S. Air Force
1670 Air Force Pentagon
Washington, DC 20330-1670
www.airforce.com/
Go to the U.S. Air Force's website to find out more about this branch of the military and jobs involving GPS.

Publisher's note to educators and parents: Our editors have carefully reviewed these websites to ensure that they are suitable for students. Many websites change frequently, however, and we cannot guarantee that a site's future contents will continue to meet our high standards of quality and educational value. Be advised that students should be closely supervised whenever they access the internet.

INDEX

ABOUT THE AUTHOR

Tanner Billings is a writer living in New York. He enjoys studying the history of the military and keeping up with modern military technologies. When he is not writing, he enjoys hiking and taking trips to national parks with his family and dogs.